MICE
of the British Isles

MICHAEL LEACH

CONTENTS
Introduction 2
Identification 4
Food 7
Breeding 11
Behaviour 18
Ecology 21
Further reading 24

Cover: *Harvest mice (Micromys minutus)*.

Series editor: Jim Flegg.

Copyright © 1990 by Michael Leach. First published 1990.
Number 54 in the Shire Natural History series. ISBN 0 7478 0056 1.
All rights reserved. No part of this publication may be reproduced or transmitted in any form or by any means, electronic or mechanical, including photocopy, recording, or any information storage and retrieval system, without permission in writing from the publishers, Shire Publications Ltd, Cromwell House, Church Street, Princes Risborough, Buckinghamshire HP17 9AJ, UK.

Printed in Great Britain by C. I. Thomas & Sons (Haverfordwest) Ltd, Press Buildings, Merlins Bridge, Haverfordwest, Dyfed SA61 1XF.

Introduction

This book covers the house mouse *(Mus musculus)*, the wood mouse *(Apodemus sylvaticus)*, the yellow-necked mouse *(Apodemus flavicollis)* and the harvest mouse *(Micromys minutus)*. Along with the black and brown rats, these species make up the British representatives of the sub-family Murinae, of which there are a total of 408 species throughout the world. The Murinae (which simply means mouse-like) belong to the order Rodentia (derived from the Latin verb *rodere*, to gnaw). With 1702 species, rodents make up almost 40 per cent of the mammal species of the world. They can survive in an incredibly diverse range of habitats, living off an equally diverse diet. In varying forms, rodents can be found everywhere on earth with the exception of Antarctica. The majority are adaptable, intelligent and highly opportunistic. After man, rodents are the most widespread and successful group of mammals on earth. There are several reasons for this success. Rodents are a relatively young group in terms of evolution — they have been here only some thirty million years. Effectively, they are still exploring the endless possibilities of natural selection. This process is speeded considerably by their high breeding rate, which enables changes to be effected rapidly. However, it is diet and behaviour that are the main areas of evolutionary alterations. Despite the variation in size, from the South American capybara weighing 66 kg (145 pounds) to the tiny harvest mouse weighing just 6 grams, their basic physiology remains fairly constant.

The rodent's prime characteristic is its teeth, particularly the incisors. This single pair of sharp teeth is essential to the animal's survival. They are capable of biting through the hardest shell or husk and gnawing through dense obstacles such as wood or even concrete. The constant use of incisors on hard substances causes considerable wear, so these teeth have open roots, allowing them to keep growing throughout the life of the animal. They are also self-sharpening, the razor-like front edge being kept sharp by the action of the opposing tooth. The front side of each incisor is coated with enamel, while the rest of the tooth is ordinary dentine. During gnawing and eating, the hard enamel at the front wears much more slowly than the softer dentine behind. This unequal wear helps to retain a chisel-like cutting edge. Between the incisors and molars is a gap called the diastema. When a rodent gnaws, its lips and cheeks are sucked into the diastema, closing off the back of the mouth and sealing it from particles such as dust and splinters. Each of the four British mouse species has twelve molars, two rows of three in the upper and lower jaws.

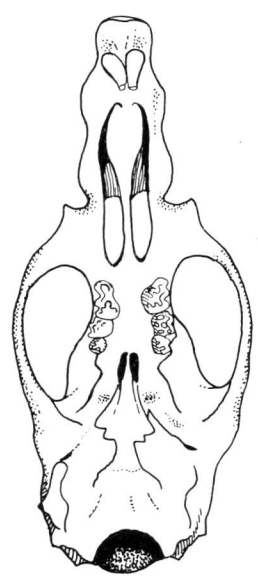

1. *Skull and upper jaw of wood mouse.*

Rodents are split into three groups: cavy-like (Caviomorpha), squirrel-like (Sciuromorpha) and mouse-like (Myomorpha). This grouping is based on the different arrangement of jaw muscles resulting in a variety of skull shapes. In mice a complex and powerful muscle known as the masseter operates the bottom jaw in a vertical plane, closing it on the upper jaw. But it simultaneously pulls the lower jaw forward, so the upper and lower incisors meet and provide the unique gnawing action. The lower jaw can be freely rotated on both a forward and a sideways axis.

The senses of mice are very well developed, as would be expected in a group that features in the diet of so many predators. The sense of smell is keen, and hearing is particularly acute. Harvest mice have small ears but the other three British species have large ears that act in much the same way as radar dishes. Each moves independently and can be swivelled through 180 degrees towards relevant sound sources. They are capable of detecting high frequencies. All four mice use ultrasound to communicate before weaning and later as adults at times of stress or excitement. In low-light conditions hearing is probably the most important sense when confronted with potential predators. All rodents use their long facial whiskers, or vibrissae, to some extent. In complete darkness these are used as 'feelers' and they are also sensitive to vibrations.

One of a rodent's most versatile and important features is its highly manipulative forefeet. With five fingers on each, they are well adapted for climbing and feeding. Mice have the ability to pick up small items, turn them over and thoroughly investigate them. This helps greatly in feeding as it enables the mice to pick out small edible fragments and even clean them. All four species are accomplished climbers and are capable of hanging upside down, holding on with just two feet.

All rodents have a prodigious breeding potential; left undisturbed, mice would quickly overrun the earth. However, the population is tightly controlled by the availability of food and a host of predators. The list of mice-eating carnivores is long, ranging from the common toad and the blackbird, both of which have been known to eat harvest mice, up to the fox and the buzzard, which prey on the larger species. Mice are capable of living up to five years in ideal captive conditions but the lifespan of their wild counterparts is a tiny fraction of this. Depending upon local conditions, the mean life expectancy of a mouse would be around eighteen weeks. Few wild mice reach the age of six months and individuals older than one year are very rare indeed.

None of the British mice hibernates: rather, as their energy requirements are higher because of the cold and finding food is more difficult, they tend to be more active in winter. In temperate climates winter is a major factor in the control of the rodent population. The lowest number of wood mice occurs in

2. *A house mouse using its forefeet to manipulate food.*

3

March and early April, after food left from the previous autumn harvest is exhausted and before the new growing season begins.

Identification

HOUSE MOUSE

This is the most widespread and visible of the British mice. House mice can be found throughout the mainland of Britain and Ireland, and on most of the inhabited offshore islands. House mice are grey-brown, a colour very different from that of the other species. House mice are one of the most genetically variable of all mammals. Individuals are regularly recorded with very light undersides or particularly short tails. Albinism and melanism are known throughout their range. In captive conditions, through selective breeding, house mice can be produced in varieties of grey, brown, fawn and black and any permutation of these colours.

The eyes and ears of the house mouse are much smaller than those of wood or yellow-necked mice. Its tail is thicker and appears more scaly than those of other species. One infallible identification aid is the notched upper incisor of the house mouse. This notch is clearly seen as a deep indentation at the back of the incisor just above the biting edge. The combined head and body length of an adult varies between 70 mm and 90 mm. The tail length is approximately 80 per cent of the body and head length. The ear length is 12-15 mm and the hind foot is 16-19 mm long. The body weight of adults ranges from 7.5 to 22 grams for the female, 10 to 28 grams for the male.

Even when the animals themselves are not seen, their presence is obvious. Footprints, well worn paths, gnawing damage to wood or any other hard surface and nibbled food are all signs that house mice are in residence. These animals have oily fur, and black smudges can usually be seen close to regular runs. Whereas the other three species are almost odour-free, house mice have a distinctive stale and acrid smell which permeates the surroundings of well established populations. Droppings are black and cylindrical, around 6 mm long and 2 mm in diameter. These can be widely scattered or in common latrine areas where large numbers are present.

3. *Cross-section of upper incisor of harvest mouse (A) and house mouse (B), showing the characteristic notch.*

4. *House mice contaminate more food than they consume.*

5. *House mice will occupy any suitable habitat.*

6. *The coat colour of house mice is very variable. Albinos are not uncommon.*

WOOD MOUSE

Although wood mice are not seen as frequently as house mice, they are the most common rodents in Britain. After man they are also the most numerous mammal in the country. They are found almost everywhere apart from densely populated urban areas and open countryside which has no ground cover, such as high mountains. There are several distinguishing characteristics that help separate wood mice from house mice. The coat of a wood mouse has a definite red tinge to it, as opposed to the grey colour of a house mouse. The total head and body length of an adult wood mouse is between 80 and 110 mm, with the tail measuring about the same. This gives it the alternative name of long-tailed field mouse. Ear length is 15-17 mm and the hind foot is 20-23 mm long. Body weight is variable and can depend on the season. In winter the average weight is 18 grams and in summer it goes up to 37 grams.

As wood mice do not enter buildings as frequently as house mice they often go unnoticed, even when living in domestic gardens. Feeding signs are probably the most obvious indications of their presence. Soft fruits like hips and haws are opened up so that the mouse can eat the seed inside; the flesh is simply left. Harder foods like hazelnuts are opened up by gnawing. Wood mice leave neat, slightly oval holes with radiating toothmarks around the bevelled edge. Softer foods, such as chestnuts, clearly show teeth scrapes. Droppings are a little larger than those of the house mouse. Mice are usually not heavy enough to leave distinct decipherable footprints in anything other than light dust. In snow, however, the wood mouse's presence is recognisable by the stripe track left by a dragging tail. Entrance holes to underground runs are usually quite exposed. They are circular with a diameter of around 2 cm, but this can vary depending on the soil structure and amount of use.

YELLOW-NECKED MOUSE

Mouse identification becomes difficult with this species. Even experts find it hard to differentiate between yellow-necked mice and wood mice. They have the same basic coat colour, though yellow-necked mice tend to be a little lighter on the underside. As the name suggests, the prime difference is a yellow collar which forms a bib on the animal's chest, joining the darker fur on each shoulder. The yellow neck varies considerably from animal to animal and can often be indistinct and difficult to see in the field. On average adult yellow-necked mice are 1½ times heavier than wood mice, with a corresponding 10 per cent increase in body and tail length. It is sometimes possible to identify dead specimens from their skulls by measuring the depth (anterior-posterior) of the upper incisors. This requires precise techniques as the measurement for wood mice ranges from 1.1 to 1.3 mm, while the range for a yellow-necked mouse is from 1.45 to 1.65 mm, but there are many records of incisors falling in between these values, making absolute identification impossible.

The distribution of yellow-necked mice is of limited help in distinguishing the species. The yellow-necked mouse is restricted to the south-east of England and an area on the Welsh border and around the Bristol Channel. Both yellow-necked and wood mice can be found in these places, often sharing the same habitat and living alongside each other.

HARVEST MOUSE

There should be no problem distinguishing this mouse from the other species, simply because of its size. The harvest mouse is the smallest rodent in Europe, weighing just 6 grams when fully grown (pregnant females can be as heavy as 15 grams). Combined body and head length is 50-70 mm and the tail measures 50-60 mm. The head of a harvest mouse is more reminiscent of a vole than of a mouse. The muzzle is blunt, the ears are short, well rounded and hairy, and the eyes are much smaller than those of any other British mouse. Young harvest mice have grey-brown fur which changes to a rich golden-brown in the adults. This colour is very different from that of the other species of mice.

One infallible way of identifying a live harvest mouse is to watch its tail. It is the only British animal with a truly prehensile tail that can be used as a fifth limb. Other mice and rats are capable of using the tail for balance and crude support by wrapping it loosely around twigs, but it cannot take the full weight of the animal. Only about 40 per cent of the harvest mouse's tail, the end portion, can be used to apply a precision grip. This part is wrapped around any available stem or twig and can act either as a brake when climbing or as an anchor when feeding, leaving the forelimbs completely free to find and manipulate food. The tip is sensitive and can feel its way about with the same skill as the other limbs. The prehensile tail, combined with an uncanny sense of balance, makes the harvest mouse the most skilful and agile member of the entire group.

Because of their small size harvest mice face a larger array of predators than the other mice. As a result they are very wary and react quickly to potential danger. In addition to sight, hearing and smell, harvest mice have a remarkable ability to sense vibrations. Movements are detected through the soles of their feet, giving the mouse an accurate picture of hidden activity. Larger animals can be sensed by vibrations passing through the ground and up the plant on which the mouse is feeding.

Harvest mice are a little observed species that was not recognised in Britain until the mid eighteenth century, when it was keenly studied by the naturalist Gilbert White. The lack of knowledge reflects the fact that there were few observers rather than a scarcity of the animals themselves. Harvest mice have a wide distribution in England, being totally absent only from the north-west and parts of the Midlands. Scotland and Wales have some records but they are few and far between. There are few indications to show the presence of harvest mice, other than nests (see 'Behaviour'), as feeding signs are too small to be noticed.

Food

Rodents have a high metabolic rate which requires a substantial amount of energy to sustain. On average they consume the equivalent of 10 per cent of their own body weight each day. The harvest mouse often increases this to 30 per cent. The exact make-up of the animal's diet is dependent upon local conditions but mice are surprisingly undiscriminating in their choice of food. In all species the basic diet consists of seeds, fruit, nuts and young shoots. However, mice are far more omnivorous than they first appear. Caterpillars, snails, flies and moths are just some of the invertebrate food that is readily taken by all four species. They will even nibble at old bones and, occasionally, consume the bodies of mice that have died naturally. Much of their food intake is converted to energy used to heat the body, and therefore they require more in wet or cold conditions.

As the major part of the diet of mice is made up of vegetation, their digestive system is well suited to handling large

7. *House mice are opportunistic feeders that readily seize the chance to exploit a new food source.*

8. *Even at close quarters it is difficult to distinguish this yellow-necked mouse from its smaller relative, the wood mouse.*

amounts of cellulose. The major anatomical adaptation is the large caecum containing a mass of bacteria, which is capable of breaking down the tough cellulose into more easily digestible carbohydrates. These cannot be absorbed in the caecum; full digestion is achieved in a very convoluted way. Consumed food is first softened in the stomach. It then goes into the caecum, where the bacteria break it down. The food is then passed out through the anus and is immediately re-eaten. The first faecal pellets are soft and on their second passage through the digestive system the carbohydrates can be absorbed easily. On their second appearance the pellets are hard and dry. This process is known as refection and is very efficient as 80 per cent of the potential energy contained in a rodent's diet is digested and used.

As the diet and approach to food of the various species of mice are different each will be dealt with separately.

HOUSE MOUSE

Because of their proximity to man, the diet of an average house mouse can be said to be basically the same as ours. In a 'natural' habitat, away from human habitation, house mice have a feeding behaviour similar to that of wood mice. But their ability to share our buildings means that they can also share our food. House mice will eat almost anything that can be found in shops, warehouses, barns or larders: bread, grain, fruit, flour, rice and most other foodstuffs. They have a remarkably sweet tooth and have often been known to go to great lengths to reach special treats such as chocolate and biscuits.

However, they are also willing to eat the most unlikely substances when local conditions offer no other choice. Soap, wood, paper and plaster are just some of the apparently inedible materials that have been eaten by house mice. These animals can survive with very little water, even when their food has a low moisture

9. *The blackened tip of this wood mouse's tail shows that the skin has been shed in response to an attack from a predator.*
10. *Hazelnuts eaten by a wood mouse.*

content. They can live for a considerable time under these conditions, but lack of water reduces their fertility. House mice are notorious nibblers. They do not feed continuously on one item; instead they feed for a short time and move on. Faced with a packet of dried peas, a house mouse would nibble parts of half a dozen instead of eating two whole ones. Little more can be said about the diet of house mice as they are one of the world's supreme opportunists. They are non-specialist and will take advantage of almost any available food supply. Their feeding strategy enables them to exploit a wide range of habitats.

WOOD MOUSE

Under similar environmental conditions, as far as we know, the diet of wood mice is identical to that of yellow-necked mice. In the case of all non-specialist feeders, diet is mainly controlled by habitat. In other words, the mice will eat whatever is available, within reason. In woodlands, seeds make up the bulk of the diet, particularly in autumn and winter. These are supplemented by nuts, fruit and other vegetable matter. This pattern changes in spring and early summer, when there is a sudden abundance of insect life in the form of caterpillars and other larvae. Both wood mice and yellow-necked mice regularly eat snails, which they reach by biting through the shell. This seems to be an acquired skill as older adults are more adept than young animals.

On arable land the diet revolves around the sown crop, that is wheat, barley, corn and so on. Although modern harvesting techniques are highly efficient, enough grain is left to support a healthy mouse population long after the crop is gone. The mice do not only take seed crops, they will readily feed off root crops such as potatoes and sugar beet. In large areas of monoculture there are always a small number of vigorous weeds to be found, no matter how well sprayed the crop may be. The weed flowers and seeds add variety to the mouse diet, which is also supplemented by animal matter. Field-dwelling mice eat more invertebrates than mice living in any other habitat. These include earthworms, insects, spiders and woodlice. Consumption of invertebrates is at its peak in winter and early summer when the available grain supplies are at their lowest.

The most varied diet is probably enjoyed by mice living in domestic gardens. A well stocked garden offers a much wider array of edible plants than an arable field or average hedgerow. Bulbs, soft fruit, peas, beans, root vegetables and young shoots are examples of the potential delicacies. Mice can be some of the most destructive vertebrate pests in a garden and their secretive nocturnal behaviour makes them difficult to remove.

HARVEST MOUSE

Little is known about the exact eating habits of wild harvest mice. They are difficult to observe in the field and leave very few clues after feeding. Most research into their diet has been carried out on captive mice. This is done by offering them the same variety of vegetation and animal matter that they would encounter in their natural environment. This is not an infallible method but it does give a good indication of their diet in the wild. The diet varies with the season and habitat. In autumn, winter and spring they feed on cereal grain and the seeds of plants such as sedges, grasses, reeds and hogweed. In late spring and summer many move to young shoots of field crops and grasses. Harvest mice also eat flowers, preferring petals and nectar. In autumn they feed off hedgerow fruits such as blackberries, hips and haws.

Throughout the year harvest mice prey enthusiastically on invertebrates. Caterpillars, flies, crickets, ladybirds, harvestmen and many other species feature prominently in their diet. Harvest mice are the most carnivorous of all British mice and they do not confine their predatory habits to invertebrates. There have been many observations of harvest mice in captivity killing and eating each other. There is little evidence of this in the wild, however, possibly because physical evidence would not last long and direct observation would be almost impossible. Being skilled climbers, harvest mice can easily reach birds' nests and they have been known to eat eggs, and occasionally young chicks, when they have the chance. Being so small, harvest mice cause very little damage to standing crops, even when they are present in large numbers. Harvest mice can have a beneficial effect for farmers as they prey on invertebrate pests such as aphids.

Breeding

Rabbits are normally used as the yardstick for reproductive power, but they are poor performers compared with mice. Rodents are the most prolific group of mammals. They breed quickly, at almost any time of the year, and start at a very early age. Mice need to reproduce rapidly to maintain numbers. Their mortality rate is very high, which is only to be expected in a group that features prominently in the diet of almost every British carnivore.

Although their breeding potential is impressive, no wild mouse ever comes near to achieving it — they do not live long enough. Under ideal conditions a female wood mouse could produce eighty young, in up to ten litters in one season. Like all the mice, she has post-partum oestrus, which means that she can conceive within hours of giving birth. Wild mice are often pregnant while still suckling a previous litter. No ecological system could withstand this population growth and fortunately it does not have to. With the exception of house mice, few mice live long enough to produce more than two or three litters. To be a viable, expanding group, mice have to compress their breeding into a very short period of time.

The number of offspring also reflects the low survival expectation. Mice can have up to ten young in a litter, although the average is around six. Litter size is greatly controlled by environmental conditions. If the weather suddenly deteriorates or an important food source disappears, the average litter size drops and sometimes breeding stops completely. On the other hand, in August and September, when seeds and fruit are plentiful, the litter size reaches its maximum for most animals. This strategy ensures that resources are not wasted when there is little chance of survival for the young, but when food is available in late summer there are large numbers of mice around to take advantage of it.

It used to be thought that the reproduction rate of prey species, such as mice, responded to a decrease in the numbers of predators. This often happened on shooting estates where hawks, stoats, weasels and other 'vermin' were destroyed on a large scale. It was soon noticed that the rodent population increased immediately. This was not through any change in breeding patterns, but because fewer mice were being eaten. The removal of the natural controlling effect of predators allowed the mice to live longer and therefore produce more young.

The length of the breeding season varies considerably. In the south of England, where winters are mild and, occasionally, non-existent, mice can breed from February to November. Further north, where the winter conditions are more severe, the season can be shortened to last from April to late September.

Mice do not form pair-bonds as mating is carried out on an opportunistic promiscuous basis. When a female becomes sexually receptive, her scent attracts all local males. The tell-tale smell is carried by pheromones, chemical signals, that appear in the urine of female mice shortly before they reach the peak of receptivity. Female pheromones effect male behaviour but there is also a reverse condition. The presence of a strange adult male in her territory will often bring a female into breeding condition much earlier than her cycle indicates. In evolutionary terms each female needs to produce as many offspring as possible and one way of achieving this is to synchronise oestrus with the availability of mates. In high-density areas males will fight amongst themselves for access. This means that only the strongest, most dominant individuals will get the chance to mate. In less populated habitats the female will mate with the first available

11. (above left). *The prehensile tip of the harvest mouse's tail immediately grips any nearby twig when at rest.*

12. (above right). *Harvest mice will make full use of any suitable food supply within their territory, including soft fruit that grows high in hedgerows.*

13. (below left). *In early winter harvest mice eat the seeds remaining from autumn.*

14. (below right). *The nest of a harvest mouse is the most complex structure built by any British mammal.*

15. (above). *The location of house mice nursery nests is sometimes given away by the squeaks of the babies.*

16. (below left). *A female house mouse grooming her ten-hour-old offspring inside a nest made of newspaper, carpet and grass.*

17. (below right). *Food put out for birds is regularly taken by house mice, which are remarkably agile when necessary.*

caller.

Mating is brief and can take place many times over a period of five or six hours. This is the only role played by the male during reproduction. Once the female loses her sexual appeal, his interest wanes and he wanders off. In times of stress, such as predator attacks or severe hunger, embryos can be re-absorbed into the body of the female.

When young mice are born they are pink, naked, blind and helpless. The average sex ratio appears to be 1:1, or very close, in all recorded litters. Mice usually give birth with little fuss and it is over in seconds, with around a one-minute interval between each delivery. The female immediately eats the afterbirth and cleans the babies by licking them.

HOUSE MOUSE

These animals are highly productive, even by mouse standards. Inside the shelter of human buildings they will breed throughout the year, as long as there is a good food supply. They are capable of reproduction in conditions that would deter other species. House mice will breed in very low temperatures and total darkness. Mating takes place when the females reach 7.5 grams and the males 10.0 grams. Gestation lasts nineteen or twenty days and the birth can take place anywhere. Nursery nests have been found in old saucepans, inside furniture, beneath floorboards, in rolls of material and in any other site where the female feels safe. House mice living away from humans build nests similar to wood mice, only they do not usually venture above ground level. The nest will be lined with any substance found nearby. One of the problems of having mice breeding indoors is that they collect soft materials when nest building. Carpet pile, furniture stuffing and paper are some of the most likely choices.

At birth the young mice weigh just 1 gram, they are blind, deaf and completely hairless apart from short vibrissae. At one week old fur begins to grow. A week later teeth appear, while eyes and ears start to function properly. Young are weaned at around eighteen days old. House mice are more gregarious than other mice and breeding females are willing to share nests when suitable sites are few or the population is fairly high. In these communal crèches suckling becomes a free-for-all and females will feed any hungry young mouse. The babies cling on to the nipples so tightly that they can be carried away when the adults leave the nest, but they soon drop off and crawl back inside.

WOOD MOUSE AND YELLOW-NECKED MOUSE

With one or two minor differences, yellow-necked mice are believed to have a similar breeding pattern to wood mice. The litters of yellow-necked mice are slightly smaller than those of wood mice, with an average of 5.04 compared with 5.5 for wood mice. Female yellow-necked mice produce fewer litters because their breeding season is slightly shorter than that of their smaller relatives. Females become fertile when they reach a weight of 12 grams and males at 15 grams. Mating is usually confined to a dominant male and any receptive female in his territory, but subordinate males are quick to step in if they see an opportunity. Before mating the male produces a sequence of ultrasounds, inaudible to the human ear, probably to placate the female.

The gestation period is 25 or 26 days and the litter is born at night in a specially chosen nest. The nursery may be in a subterranean tunnel system or above ground in hollow logs, dense undergrowth or even in bird nest-boxes. The nest is lined with any soft material that is locally available, such as shredded leaves, grass or feathers. At birth the young weigh 1-2 grams and their development follows a similar pattern to that of house mice. Males tend to grow more quickly than females. In the long term this ensures that the males reach breeding weight first and when the females become receptive fertilisation can take place immediately. Teeth first appear at thirteen days, eyes open at about fourteen days and the mice are fully weaned at eighteen days. Any litter born at the peak of the breeding season will be driven out of occupied territories. Dominant males are particularly aggressive:

18. *Underground tunnel complex of wood mouse. The design and tunnel depth varies with local soil conditions.*

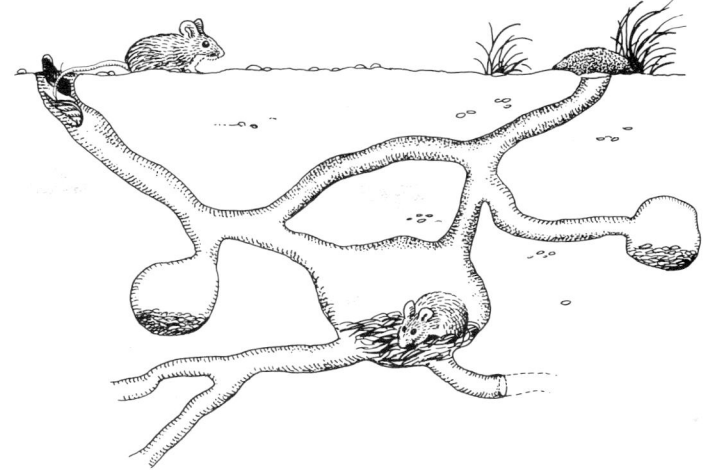

19. *(below). Young wood mice are blind and totally helpless.*

20. *Wood mice are capable of climbing high into trees to reach food. If threatened they simply drop to the ground.*

21. *During the winter months wood mice can sometimes be attracted to food intentionally left out by observers.*

22. *When first emerging from a run, wood mice are extremely wary. They always smell the air before venturing into the open.*

23. *All mice groom fastidiously every day, like this wood mouse. In times of stress they often wash for a few seconds before taking action.*

they will chase and sometimes kill juvenile mice, but this activity is much less obvious in winter. Wood mice moult at any time between five and nine weeks old, when their flanks acquire the yellow tinge of adults.

HARVEST MOUSE

Female harvest mice do not start to breed until they weigh at least 6 grams. This usually happens at about seven weeks old. It is the weight that is important: if conditions are poor, the mouse may be older before it is sexually mature. Males become mature a little earlier. Mating is swift and often aggressive, resulting in the female turning on the male and chasing him off. Once pregnant, female harvest mice build nurseries known as aerial nests. These are spheres with a diameter of about 10 cm, built anything up to 1.5 metres above ground level in live grass, cereal or reeds (for nest building, see 'Behaviour'). A nest is used only once — each successive litter has a new nursery. The gestation period is around eighteen days and during this time the weight of the female doubles. The birth usually takes place at night; the young mice weigh between 0.7 and 1.0 grams and are 15-22 mm long. Harvest mice develop more rapidly than the other species. Their weight increases at the rate of 15 per cent a day.

The young mice lie together for warmth as the female does not stay with them unless she is suckling them. After feeding she washes the babies and eats their droppings. At birth the young appear dry and scaly but within 48 hours fur starts to appear. When one week old, they start to groom themselves and shortly after this their eyes open and begin to function, as do their ears. This coincides with the appearance of teeth. When they are nine days old the female introduces them to their first solid food. This is in the form of chewed seeds brought in by the mother. When twelve days old, the adolescent mice leave the nest and start to explore but they return home to sleep. Three days later they become independent. If the female is expecting another litter, she is likely to drive away the members of the previous one.

At first the young mice are a dull brown colour but they quickly enter a moult. This begins about four weeks after their birth and can take anything from 35 to 120 days to complete. The process begins on the back of the animal and spreads forwards and sideways. It can stop at any stage for a variety of reasons and then the mouse has two distinct colours, the dull brown of the adolescent and the golden brown of an adult.

Behaviour

HOUSE MOUSE

These mice are mainly nocturnal but they will forage at any time if left undisturbed. House mice have been described as 'animal weeds', that is they are found in unwanted places, are irritating and difficult to remove. The behaviour of these mice is very adaptable. They are cautious but show enthusiasm for exploration and colonisation. Any new object inside their territory is treated with great suspicion at first, but this is followed by a complete investigation. House mice are probably the most inquisitive members of this group. Seasons have no effect on animals living indoors but there are some colonies that migrate outside in summer and back inside during the winter.

Each colony of house mice has a group scent which identifies all members. When meeting, the mice sniff each other briefly; any stranger is quickly identified and may be attacked. Aggression is accompanied by high-pitched squeaks that can easily be heard by humans. They run in short bursts, stopping to scent the air occasionally. They prefer to keep close to objects such as walls and are much more wary when they venture into open spaces. When handled by humans they are likely to bite but do not shed the tail skin in the same way as the wood mouse does.

The house mouse is not a native of Britain. It was introduced from mainland Europe around 1000 BC. The species probably originated in Asia and increased its range through association with man.

WOOD MOUSE AND YELLOW-NECKED MOUSE

Both these species are totally nocturnal and they even show a marked dislike for strong sunlight. Males are noticeably more active than females. Their movements are quick and agile, made up of short runs and much longer jumps. These mice can perform impressive leaps when trying to avoid predators. When wary, their movements are slow and the nose is kept high. They occasionally stop and stand up on their hind legs to look around and scent the air.

If predators are detected, the mice dart back to one of the entrances to the tunnel complex. They find it by a combination of sight, smell and use of the earth's magnetic field. This last factor helps them locate bolt-holes quickly on very dark, moonless nights. The tunnel systems are variable in design and size. They are often found under natural cover such as hedgerows and tree roots but can be found in open ground. Each system has several entrances, usually radiating in every direction. Inside are nesting chambers, lined with whatever soft materials are locally available, and food stores. Here the mice store any food that is abundant in their area. Nuts are useful in providing food for overwintering. The entrances to tunnels are often blocked with twigs or leaves to disguise them.

These mice do not hibernate but in winter they are known to experience hypothermia. In cold conditions the body temperature will occasionally drop to the point where normal functions are impaired. This can result in death directly through heat loss or indirectly through hunger or predation. In cold weather, when sexual competition is non-existent, they will sometimes sleep communally for warmth. The social hierarchy is not confined to breeding as dominant mice also have prior right to prime burrow sites and preferred food. For much of the time mice probably avoid each other to prevent any possible aggressive behaviour. Communication is very limited. Individual mice recognise each other by body scent or droppings. Audible squeaks are given out if the mice are frightened and, by the males, before and during mating.

HARVEST MOUSE

This species does not follow any rigid time patterns. Its periods of activity vary considerably, but it does show a slight preference for a nocturnal existence in summer, shifting to a diurnal one in win-

24. Wood mice and yellow-necked mice use powerful jumps to escape predators such as weasels.

25. *A traditional method of dealing with unwanted mice: however, chocolate is a much more effective bait than cheese.*

26. *This wood mouse wandered into a discarded bottle to drink and cannot get out.*

ter. This ensures that the mice avoid leaving the nest during the cold winter nights. Whatever the timing pattern, these mice are active in bursts of three to four hours, punctuated by shorter periods of rest. The seasons greatly influence the activity patterns as the small size of the mice means that they can lose heat very quickly. In winter they move down to ground level for various reasons. The most obvious one is that much of the vegetation dies back, but it is more complex than that. Life at ground level is less exposed to the elements. The mice can find shelter against winter wind and rain amongst the bases of both dead and living plants. When it snows the mice simply stay beneath the drifts; these act as an efficient insulation against cold while the mice forage, safe from most predators, for seeds at ground level. Wherever possible, they will often move into deserted underground tunnel systems.

Whatever the season, harvest mice build nests. All individuals build non-breeding shelter nests, which are usually found at the bottom of plant stems but can also be found under rocks, in hollow logs, hay bales or even abandoned birds' nests. Non-breeding nests are made of shredded grass leaves that are loosely woven together. They are made very quickly and do not last long. Breeding nests are completely different. They are situated much further off the ground and are built to last a lot longer. They are constructed by using the stalks of living grass as girders around the outside and then nearby leaves are shredded, but not removed from the stem, and woven tightly into a sphere. Once the basic shape is made, the mouse moves inside and pulls more shredded leaves through the wall. These are then woven into the inside fabric. All of the leaves are left attached to stems and the nest is left suspended between, and connected to, several plants. It is then lined inside with thistledown or finely chewed grass.

Harvest mice are not a sociable species and often fight when they meet. The first encounter consists of both animals sitting on their haunches with their heads high in the air, exposing the incisor teeth. They then 'box' and chatter furiously. Eventually one withdraws and runs off. It is usually chased by the remaining mouse, who bites at the rear of the subordinate. This can result in the loss of tails or even hindfeet.

Ecology

The survival strategy of many rodents means that they can exploit a wide variety of habitats, the success of which is based upon the presence of three basic factors: acceptable climate, suitable food supply, and safe sites for shelter or breeding. The behaviour of each species, such as the type of cover they prefer and the extent to which they accept human disturbance, also affects the choice of home territory. When the population density is high, young or subordinate mice are often forced to occupy areas that are less than ideal because the best sites are already taken. Survival is always more precarious in anything other than optimum conditions.

HOUSE MOUSE

This animal is so indiscriminate in its choice that there is no need to look closely at its habitat. It will live virtually anywhere. Despite its name, this mouse does not live only in buildings. There are large numbers in woodland, hedgerows and arable fields that never come into close contact with humans. They have also been known to live in cold stores, where meat is kept at temperatures barely above zero. To compensate for these conditions, cold-store mice grow an extra-thick coat to keep warm. The home range of indoor mice is controlled by the availability of food, but a typical territory would cover about 5 square metres. Outdoor mice often have less defined ranges and can sometimes be semi-nomadic.

In good conditions house mouse populations can become completely out of hand. Plagues of mice are an occasional problem wherever the species occurs, although they are not as common in Britain as elsewhere. One night an Australian farmer found 28,000 dead mice on his verandah after laying poisoned bait. The population grows to whatever number can be supported by the environment. In warehouses and grain stores, where the food supply is endless and there are no natural predators, numbers can become huge. The damage caused by house mice can be considerable and costly. Stored food is partially eaten but an even bigger quantity is spoiled by the animals' droppings, which can transmit a variety of ailments to man, including leptospirosis and lymphocytic choriomeningitis. Habitual gnawing of hard surfaces creates potential fire hazards when the mice chew through electric cables.

With so many disadvantages associated with resident mice, large sums of money have been spent in trying to remove them. Simple traps are effective but limited in the numbers that they catch. There is evidence to show that mice begin to recognise these as dangerous and avoid them after a time. The traditional cat acts as a deterrent but little more. Mice have systems of boltholes and are quick to see predators. Cats tend to take the slower or injured individuals and rarely make much impression on established populations. Poisoning is still the most efficient control technique. At first this consisted in leaving out tempting food laced liberally with simple substances such as strychnine or cyanide. These are fast-acting agents and mice soon recognised the danger and avoided them. Anti-coagulation compounds such as warfarin were then introduced. These block the clotting process in blood, causing the affected animal to bleed to death, internally or externally. But, before long, mice developed a genetic resistance to this. A second wave of more powerful anticoagulants is now being used with a high success level. It is possible that in time the mice will be able to resist these too. It would be a mistake to underestimate the adaptability and tenacity of these animals.

WOOD MOUSE

The common names of animals are often misleading, as is the case with the wood mouse. This animal is also known as the long-tailed field mouse, but even this does not give an accurate idea of the habitat of the animal. This species will occupy any suitable niche from low heather moorland and open fields to occupied buildings and deciduous woodland. Providing the basic essentials are available, wood mice show no preference for any specific habitat type. However, they do seem to dislike high land above the tree line, open grassland and mature conifer plantations. Wood mice can be found in cereal fields, where cover is provided by their tunnel complex. Food is abundant and their foraging range is relatively small, so that feeding mice do not have to wander far from the safety of underground runs. They are frequently found in urban areas, living in parks, domestic gardens and wasteland. These animals like plenty of cover and they take full advantage of man-made habitats such as hedgerows and dry-stone walls. These offer ideal breeding sites and boltholes and also act as bases for foraging trips into nearby feeding sites. Walls and hedges form a network of interconnecting linear habitats that shelter mice in areas where they could not otherwise exist. They are wildlife corridors which provide a perfect opportunity for travel and colonisation of new habitats.

The size of the home range of a wood mouse is dependent on many factors. Where the food supply is varied and readily available, the mice tend to occupy small areas. But in less favoured sites the mice will have to venture further to reach food. Winter ranges are smaller than summer ranges, probably because there is less cover and the mice are therefore more vulnerable. One survey in woodland showed that male ranges, at 2250 square metres, were larger than the female ranges of 1817 square metres. From weaning, an individual's home range slowly increases until it reaches sexual maturity.

The population density of wood mice is just as variable as that of the other species. In mixed woodlands at the end of the breeding season in autumn numbers can reach 100 mice per hectare. The trough occurs in March and April, when following a hard winter numbers can be so low that some populations die out. Unoccupied suitable habitats are quickly colonised from nearby populations once breeding starts again. Life expectancy of wood mice is low. From weaning an average lifespan is between eight and fourteen weeks, although in captivity wood mice have been known to live up to two years.

YELLOW-NECKED MOUSE

These are not as widespread as wood mice but they share similar habitats. They seem more wary than their smaller relatives, as yellow-necked mice are seldom found in open fields and they are not likely to colonise built-up areas in the same way. Yellow-necked mice are always found sharing their habitat with wood mice. It seems that the wood mice avoid any direct confrontation with the larger yellow-necked mice. This relationship is not yet fully understood as the two species are known to interbreed on the European mainland, but this has not yet been recorded in Britain. In most shared habitats the number of yellow-necked mice is considerably less than that of wood mice. One explanation for this is that yellow-necks have larger home ranges, thus reducing the sustainable population levels in any given area.

Yellow-necked mice are skilful climbers and spend more time feeding in trees and shrubs than the wood mice do. This skill results in yellow-necked mice entering buildings more often than the smaller wood mice. They are capable of scaling brick walls with ease, to reach open windows or even air spaces in the eaves. Resident yellow-necked mice are not as destructive as house mice, but they can cause considerable damage to stored food. The mortality rate and population structure is very similar to those of the previous species but much less research in these areas has been done with yellow-necked mice.

HARVEST MOUSE

These wary rodents are more discreet in their choice of habitat and they move about as the seasons change. They prefer areas of thick tall vegetation and, despite their name, can be found in a wide range of environments. These include reed beds, hedgerows, roadside verges, riverbanks and overgrown gardens, as well as the better known sites of cereal-sown fields. During the summer months harvest mice rely solely on grasses, reeds or sedges as the basic materials of their nests. Any chosen habitat has to contain a fair percentage of these for the mice to thrive. Under good environmental conditions the population of harvest mice can reach a density of up to 220 per hectare in early autumn. But this is a seasonal peak that is rapidly reduced by the hardships of winter. 99 per cent of mice found in autumn will be dead before the start of the following breeding season in spring. Harvest mice have a very short life expectancy. Long-term trapping studies show that only 30.6 per cent of mice survive more than six weeks after the day they were caught in a trap. The figures are quickly lowered until they show only 0.7 per cent of them survive more than thirty weeks from the original trap date. The maximum lifespan of a wild mouse would be around eighteen months, but only a tiny number would even approach this age. However, in captivity, non-breeding individuals have been known to live as long as five years.

The high mortality figure is caused in part by the fact that the habitat of the harvest mice can change rapidly and can often completely disappear. In a few hours the crop of a field can be collected by a harvesting machine. Meadows can be grazed down to earth level by a small flock of sheep. These mice live in transitory habitats that alter on a seasonal basis, even if nothing else disturbs the equilibrium. As vegetation dies down in winter, the mice can either change their behaviour and become ground-level dwellers instead of stalk-level, or they can move to somewhere else. Haystacks, barns and woodlands are a few of the places used for overwintering.

The home range of harvest mice is small. As they make full use of all three dimensions of their habitat, they have little need to cover a large surface area. The range of a male mouse in grassland would be anything from 300 to 900 square metres, with females having a slightly smaller range no larger than 800 square metres. Home ranges of individual mice can overlap to a degree, without causing much aggression.

PREDATORS

Because of their non-specific habitats and behaviour all four species of mice are liable to be taken by any one of many potential predators. There are no specialised mouse hunters in Britain, but most predators incorporate mice into their diet to some extent. The most hazardous existence is experienced by harvest mice. Their small size makes them vulnerable to attack by all of the obvious predators such as weasels and kestrels but also by less likely species such as blackbirds, toads and pheasants. As harvest mice are active during the day and the night, they are exposed to both nocturnal and diurnal predators. The most expert harvest mouse hunter is probably the barn owl. In winter and during the breeding season these birds will hunt at any time of the day or night and they share a similar habitat to harvest mice. One researcher discovered that up to 65 per cent of barn owl pellets were made up of harvest mice remains. This would be a temporarily high figure brought about by a peak in the mouse population.

The other three species face a less diverse array of predators as they are slightly larger and are active mainly at night. Long-eared owls and tawny owls are frequent predators of yellow-necked mice and wood mice. Pellets show that around 30 per cent of their diet is made up of the *Apodemus* mice. Both of these species have evolved an effective way of escaping death under some circumstances when caught by a predator. Under conditions of extreme stress they can shed the skin from the tail if it is gripped anywhere along its length apart from the base. Once this happens, the skin is not regenerated: the exposed flesh and caudal vertebrae dry out and eventually drop off.

The chance of survival from predation is highest in winter, when the number of predators is at its lowest. The highest rates of attack take place in late spring and early summer when predators and their offspring are both feeding.

Further reading

Corbet, G. B. and Southern, H. N. *Handbook of British Mammals*. Blackwell Scientific Publications, 1977.
Harris, Stephen. *Secret life of the Harvest Mouse*. Hamlyn, 1979.
MacDonald, David (editor). *Encyclopaedia of Mammals*. Guild, 1984.
Berry, R. J. (editor). *Biology of the House Mouse*. Academic Press, 1981.

USEFUL ADDRESSES

The Mammal Society, Baltic Exchange Buildings, 21 Bury Street, London EC3A 5AU.

ACKNOWLEDGEMENTS

I would like to thank Brett Westwood for the use of his excellent line drawings. Thanks are also due to Raymond and Judith Wilson for their help and advice on the text. All photographs are by the author.